Lewis Stephen Pilcher

Further Contribution to the Study of Fractures of the Inferior Extremity of the Radius

Lewis Stephen Pilcher

Further Contribution to the Study of Fractures of the Inferior Extremity of the Radius

ISBN/EAN: 9783337811211

Printed in Europe, USA, Canada, Australia, Japan

Cover: Foto ©berggeist007 / pixelio.de

More available books at **www.hansebooks.com**

FRACTURES OF THE INFERIOR EXTREMITY OF THE RADIUS. DIFFERENTIATION OF LONGITUDINAL AND TRANSVERSE FRACTURES AND THE CAUSES WHICH PRODUCE THEM.

BY L. S. PILCHER, M.D.,

BROOKLYN, N. Y.

FURTHER CONTRIBUTION TO THE STUDY OF FRACTURES OF THE INFERIOR EXTREMITY OF THE RADIUS. DIFFERENTIATION OF LONGITUDINAL AND TRANSVERSE FRACTURES AND THE CAUSES WHICH PRODUCE THEM.

BY L. S. PILCHER, M.D.

It is now two years since I had the privilege of presenting to this Society a memoir upon fractures of the radius near the wrist joint, in which I demonstrated, by a method of my own, that transverse fractures of the lower extremity of the radius were produced by a force of avulsion, conveyed through the anterior carpo-radial ligament, a truth which, though it had been fully demonstrated eighteen years previously by Le Comte in France, had been generally lost sight of in this country. I was able to add to what had been previously described as to the pathology of these injuries two new facts, viz. : the action of the internal or carpo-ulnar fasciculus of the anterior ligament of the wrist in locking the fore-arm in a state of extreme supination, as long as the backward displacement of the lower fragment of the radius, if at all great, remains unreduced; and secondly, the frequent existence of a tense pseudo-ligament formed by the stripping up of the dorsal periosteum from the upper fragment, reinforced by the dorsal annular ligament and the sheaths of the extensor tendons, which having remained untorn, and having been forcibly put upon the stretch at the moment of the backward displacement of the lower fragment, tends to finally hold the fragment in any position of entanglement into which it may have come. In this memoir I further called particular attention to the preponderance of sprain as an indication for treatment, and having discussed the special conditions presented by sprains of the wrist joint, and having shown that fractures of the inferior extremity of the radius *per se* never entailed permanent disability, but that in the results of the sprain were to be found the fruitful source of long continued, and sometimes permanent impairment, of the function of the joint, and having shown that sufficient support for the satisfactory retention of the fragments after their reduction could be afforded without the use of splints, I advocated strongly the disuse of splints in the treatment of this injury, and the use of a simple retentive bandage with massage.

These views received from the profession in general a much more favorable notice than I would have dared to expect, and from many quarters I have received assurances from surgeons and surgical teachers, of their endorsement of the views of that memoir.

In addition, my friend and colleague, Prof. Jarvis S. Wight, has done me the honor of carefully considering my memoir and of commenting upon it at length in an erudite and logical paper which appeared in the *Medical and Surgical Reporter*, of Philadelphia, in November, 1878. The conclusions which he arrived at, as inferred from his statements, are that the memoir in question does not display a full knowledge of the differing facts of the accidents of which it professed to treat, that it is disingenuous in its statements of the results of the experiments detailed, that its inferences are not sustained by the facts adduced, that its statements of anatomical conditions and of functions are incorrect, that its conclusions are erroneous, and that its doctrines are dangerous to the reputation and purses of those surgeons who may adopt them.

The high respect which I have for the professional attainments of my commentator cause any views which he should express to command at once my attention. Moreover, the authority which his position as a professed teacher of surgery gives to his words would make it impossible for me to dismiss them lightly.

Still more, the high principles, with the announcement of which he prefaces his paper, find a quick response in my own mind, and I adopt his words as my own, *verbatim et literatim :*

" The object in writing this paper is to establish scientific truth, to lay a foundation for correct practice, and to aid in defending the surgical profession against unjust suits at law for malpractice. It is safe to say that the truth is better than any man; but if the false has been made to seem to be true, and thereby one reputable surgeon has been unjustly ruined in reputation and prospect, it is also safe to say that the responsibility of his ruin rests with those who have, perhaps without intention, made the false seem true."

My friend evidently felt that it was an unpleasant duty which he had to perform, for he proceeds to say that, by the combination of various circumstances, it is imposed on him as a simple and imperative duty, a duty which is fully discharged only when he has lectured the three eldest and most renowned surgical teachers of New York City, singling them out by name, upon their dangerous course in endorsing, in any degree, these new doctrines !

I have felt that it was incumbent upon me to re-examine this whole
question with whatever of candor and capacity I could command, and
either to candidly acknowledge my errors, if such I had fallen into, and
to recant from any dangerous doctrines, or if, on the other hand, further
examination and experience should more firmly convince me of the cor-
rectness of my former positions, to do my best to convert my deceived
friend from the error of his ways, and to place him on that basis of scientific truth he holds so dear. I have not found the year and a half
which has elapsed since the publication of Prof. Wight's article any too
long to accomplish the task which I had undertaken, while it has been
long enough, I hope, 'to enable me to bring to this present discussion a
spirit of calm scientific inquiry. The Professor, after the settling of some
preliminary measurements and definitions, which indeed impress one with
the care and research which he brings to his task, opens the discussion of
the relations of the hand to the radius in forcible extension, by comparing
the hand to a lever of the first order, the radius being the fulcrum upon
which it rests. The hand constitutes the long arm, the arm to which
the power is to be applied, and the short arm is *that portion of the carpus
between the insertion of the flexor muscles* and the point of greatest pressure
on the inferior extremity of the radius, the length of which, he states,
is apparently one inch.

Here at once I am bewildered, for in all the wrists I have ever examined, I have never seen one yet in which, in extreme extension, any part
of the carpus projected one inch in front of the point of greatest pressure against the articular surface of the radius. Then again there is but
one little flexor muscle—the flexor carpi-ulnaris—attached to any part of
the carpus, either before or behind the point of greatest pressure against
the radius, while the great mass of flexor tendons sweep over the trochlear
surface afforded by the carpus, and pass forward to their insertions into
metacarpus and phalanges.

The anatomy of my critic is certainly at fault at the outset. Perhaps
his mechanics will be better. Proceeding with his original hypothesis of
the lever, he states that the ratio of power and weight force, as determined by the relative lengths of the two arms of the lever, will be as 1 :6.
And then he says: Suppose 100 pounds of extension be applied to the
power end of the hand lever, there will be a pressure of 700 pounds on
the lower extremity of the radius. Now, this hypothesis is entirely un-
tenable, for the very essence of it is that six-sevenths of this weight is
muscular contraction, and it goes without saying that the contraction of
the muscles will have been overcome long before the hypothetical amount
is reached, and the terrible results of a pressure of 700 pounds on the
lower extremity of the radius will have been averted. But even here he

has persisted in assuming that the hand lever is a lever of the first order, while really it is a mixed lever, viewed from the standpoint of the flexor muscles, being partly of the second and partly of the third order, and partaking not at all of the first order. Incorrect in anatomy, incorrect in mechanics; and these in his premises; there is a possibility that his ultimate conclusions may be incorrect also.

A series of eleven cases are next marshaled in order—two from his own experience, and seven collated from various text books. The two first are entirely irrelevant—one of them does not pretend to be a fracture of the kind under discussion, and the other presents no proof of ever having been any kind of a fracture. The seven collated ones illustrate very well the differing conditions which fractures of the lower extremity of the radius may present. Later follows a description of the experiments of Nelaton, and their repetition by Dr. Daniel Ayres, with the positive assertion ''so that M. Nelaton made a Colles fracture in the dead fore-arm, by compressing the base of the radius between two counterforces; and this on the authority of Malgaigne.''

This is his *piéce de resistance*, upon which, after all, it is apparent that he bases chiefly his ultimate conclusions. The doctor does not seem to be aware that the real character of the experiments made by Nelaton were quite different from what both that distinguished surgeon and his contemporary, Malgaigne, supposed them to have been, and that 18 years ago the error of these gentlemen was demonstrated, and their experiments shown to have proved an entirely different thing. Finally the doctor makes the issue squarely on the assertion contained in my original memoir, that the theory that fracture of the lower extremity of the radius happens because the bone, having been caught between two forces, gives way at its weakest part, is not sustained by the facts of the case, and as the conclusion to his own reasoning formulates the following theorem:

"A fracture of the base of the radius is generally caused by the reaction of the resisting surface on which the palm of the hand strikes at the time of the fall, the carpus being driven or pressed against the base of the radius."

I accept the challenge, as thrown out by the Professor, as to the facts of the case, and as a basis for argument repeat, with the addition only of the word transverse, as modifying the word fracture, my original theorem:

Transverse fracture of the radius near the wrist joint results almost invariably from falls, the force of which is received upon the palm of the hand.

The theory usually adduced in explanation of the injury resulting is that the radius, having been caught between two counter-forces, has given way at its weakest point, about one-half inch from its lower extremity.

An examination of the facts of the case does not sustain this theory.

For purposes of study these facts may be classified as follows :

First, those pertaining to the place and character of the fracture.

Second, to the anatomical structure and relations of the fractured part. Third, to the character of the violence which has been known to determine these fractures. .

In studying the place and character of fractures of the lower extremity of the radius, I have examined eleven specimens of recent fracture contained in the Museum of the New York Hospital, and one in the collection of Professor F. H. Hamilton, an analysis and study of which have not heretofore been made.

In their examination I note : First, that ten of the twelve present lines of fracture running both transverse to and parallel with the long axis of the bone ; in the greater number these vertical lines of fracture involve only the lower fragment, which is comminuted as the result ; in some these vertical lines extend as lines of fissure up into the shaft itself.

Second, that impaction of the upper in the lower is present in only one specimen—No. 123—and in this so slightly, that little force would be necessary to overcome it.

Third, that a great variation exists in the amount of the articular extremity which has been separated from the shaft.

In one specimen—No. 125—the lower fragment is composed of the tip only of the styloid process of the radius ; in a second—No. 127—the whole styloid process is broken off, the line of fracture running obliquely from its base downward to the bisecting ridge of the articular surface ; in these two specimens the dorsal lip of the articular surface of the radius has been crushed off, and a fracture of the scaphoid bone also sustained ; in a third specimen—No. 122—a yet larger portion of the articular surface is included with the fragment, the line of fracture running from the base of the styloid process to the middle of the quadrilateral articular surface for the semi-lunar bone ; in a fourth—No. 121—the whole of the carpal articular surface of the radius is carried away with the lower fragment, by a line of fracture running very obliquely from the base of the styloid process to the point of junction of the carpal, and the semi-lunar articular surfaces.

In other specimens—Nos. 123, 124, 129, 125ª, 120 of the New York Hospital collection, and the specimen belonging to Professor Hamilton—the lower fragment includes the whole of the inferior extremity of the bone, which is separated from the shaft by a line of fracture running nearly transversely at points varying from one-quarter inch to one inch above the articular margin in front, with an antero-posterior obliquity, so that on the dorsal aspect of the radius the line of fracture is somewhat higher above the articular margin than was the case upon the palmar aspect ; in the specimen—No. 124ª—the line of fracture is lowest upon the dorsal aspect of the bone where it runs along the epiphysial line about ⅜ in. above the articular edge, thence it runs obliquely upwards as it approaches the palmar aspect, where it terminates about one inch above the articular edge.

In the remaining specimen—No. 128—there is simply a longitudinal fracture running upward from the articular surface, near the ulnar margin, by which a bit of the radius containing the ulnar articular facet is split off. This was accompanied by a fracture of the scaphoid bone.

It is to be noted that all these were instances of injuries, where a great degree of violence was sustained, and other injuries of fatal character received.

As to the place and character of fractures of the lower extremity of the radius, I conclude therefore from these specimens that the fractures which do occur are of two distinct general classes—transverse and longitudinal ; that in transverse fractures the amount of the lower extremity separated from the shaft is extremely variable ; that the line of transverse fracture is rarely above one inch above the articular margin ; that it may approach it very closely, may involve it at any point, and may include only a portion of the styloid process; that longitudinal fracture may co-exist with transverse fracture, or may be present as the only lesion.

The recognition of the two distinct directions of the line of fissure which these fractures may present, carries with it the recognition of the necessity of the bone having been subjected to forces acting in two distinct directions upon the bone at the time of these injuries, one acting in the line of the axis of the bone tending to split it, a force of cleavage, one acting transversely to the axis of the bone, a cross-breaking strain.

ANATOMICAL STRUCTURE AND RELATIONS.

A brief enumeration of those prominent points connected with the structure and relations of the lower extremity of the radius which may be of importance in tracing the connection between the place and character of the fracture and the violence which has been its determining cause is here in place. The radius at its lower end becomes expanded in every direction to adapt it for the carpal articulation. This expansion is greatest anteriorly and externally, and results in the development of the massive styloid process and the prominent anterior lip. The thick and dense compact tissue which forms the walls of the shaft above becomes gradually thinner as it enters the expanded lower end, which is composed of cancellous tissue enclosed within a very thin compact shell. In front the thick compact wall descends with but little diminution in its thickness to within half an inch of the articular margin, and then shades off abruptly; behind, the compact wall shades off very gradually and terminates nearly half an inch higher up.

Upon inspection of a longitudinal section of the radius it will be seen that the cancellous tissue of the lower extremity forms a distinct wedge which is received between the diverging thick ·external compact laminæ of the shaft above. The projection of the anterior lip in front of the plane of the anterior surface of the shaft is equal to one-half the diameter of the shaft. The posterior lip is less prominent, descends slightly lower than the anterior, and, notwithstanding the posterior expansion of the shaft immediately above it, is itself on a plane slightly anterior to the plane of the posterior surface of the unexpanded shaft. The lower half or three-quarters of an inch of the radius makes with the shaft a slight angle, being bent forwards so as to bring the greater part of the articular surface in front of the axis of the shaft and to give an anterior obliquity to its surface.

The articular surface is slightly concave both from before backwards and from side to side. The first row of the carpus presents a sharply convex surface both from before backwards and from side to side, of which surface but a small portion is in contact with the radial articular surface at any one time. A considerable degree of antero-posterior motion is possible in both the mid-carpal and the radio-carpal articulations. This motion is most extensive in the latter articulation. By the sum of the two the hand can be flexed and extended nearly .to a right angle with the fore-arm without bringing undue tension upon the ligaments of these joints.

The anterior radio-carpal ligament unites the radius and carpus in front, being inserted above into the projecting anterior lip of the radius. The most dense and strong portion of this anterior ligament is a band which arises from the ridge just above the styloid process, and from the uneven depression between that ridge and the margin of the articular surface for the scaphoid bone, and passing obliquely downwards and inwards, is inserted into the semi-lunar, magnum and unciform bones; blending with the outer edge of this band is the external lateral ligament, which, arising from the extremity and forepart of the styloid process of the radius, passes to the inner and forepart of the scaphoid and the adjacent edge of the os magnum, so that its fibres pass to the forepart rather than the side of the joint, reinforce the anterior ligament and combine with it in limiting extension. That portion of the anterior ligament which arises from the eminence that surmounts the articular surface for the semi-lunar bone passes downward less obliquely and is attached to the magnum, unciform and cuneiform bones. From the anterior margin of the head of the ulna, from the styloid process of the ulna, and from the inner border of the ulna at the base of its styloid process, a third strong band arises, which passes obliquely downwards and outwards and is inserted into the forepart of the cuneiform and the adjacent edges of the unciform and semi-lunar bones; its inner edge blends with the internal lateral ligament, which also contributes to limit extension. The anterior ligament is further reinforced and its ability to limit extension increased by the flexor tendons of the fingers which lie upon it, together with their investing sheaths.

The posterior radio-carpal ligament permits a degree of flexion nearly equal to that of extension. Comparatively weak, it is powerfully reinforced by the extensor tendons, their sheaths and the posterior annular ligament.

The carpal and the metacarpal bones are joined together with such firmness that but slight motion is permitted between them. In the movements of extension and flexion at the wrist they act virtually as one bone.

THE CHARACTER OF THE VIOLENCE WHICH HAS BEEN KNOWN TO DETERMINE THESE FRACTURES.

The most frequent cause of fracture of the lower extremity of the radius is universally conceded to be a fall, the force of which has been received upon the palm of the hand. In many cases the fracture has resulted from falls, attended with but little violence, as where a person, having slipped, has thrust out his hand to break the fall. Where a great degree of violence has been sustained, as in falls from a height, this' particular lesion of the radius is usually found to be complicated with comminution and marked displacement of its lower fragment, and with stretching and more or less complete rupture of the radio and carpo-ulnar ligaments; the styloid process of the ulna may be broken off, and the head of the ulna caused to project strongly upon the antero-internal aspect of the wrist, and, in extreme cases, even may be forced through the skin.

Fractures of the lower extremity of the radius have been produced by simple forced extension of the hand.

Voillemier, in 1842, first called attention to this fact, and reported two cases exemplifying it. Occasional examples have been reported since ; to their number I can add the following three, observed by myself :

CASE I.—Delia Moore ; aged 30 years ; residing at 626 Classon Ave., Brooklyn, N. Y. In the evening of December 30th, 1878, while standing on a chair, she slipped, and, falling, caught at a table to save herself ; the fingers only of the outstretched hand reached the edge of the table and her hand was bent strongly backward ; upon recovering herself, she found that her wrist was injured. She immediately applied to Dr. John Cooper, of 33 St. James Place, who recognized a fracture of the lower ex-

9

tremity of the radius, with displacement. He at once reduced it, but sent her to me for further treatment. Upon examination, shortly afterwards, I was able to elicit distinct crepitus and to confirm the diagnosis.

CASE II.—Daniel Maloy ; aged 14 years ; residing at 42 Front Street, Brooklyn, N. Y. September 19th, 1879, while endeavoring to prevent a boat from chafing against a dock, his right forearm was so caught between the boat and the dock that the boat impinging against his elbow, and his hand pressing against the dock, the hand was forced into extreme extension. Within an hour after the accident I saw him, and he was also examined by Drs. F. H. Stuart and Geo. R. Westbrook. His right wrist presented the characteristic silver-fork deformity of fracture of the lower extremity of the radius ; the line of fracture on the outside was about three-fourths of an inch above the tip of the styloid process, and thence ran transversely about one-half an inch above the articular surface ; the lower fragment was displaced backward, without marked rotation, and was entangled upon the posterior edge of the upper fragment, the anterior edge of which was distinctly felt projecting upon the palmar surface.

CASE III.—Bernard Ettinger ; aged 16 years ; residing at 249 Atlantic Ave., Brooklyn, N. Y. September 23d, 1879, while assisting in lowering a case of tobacco down a staircase, he was compelled to support for a moment the whole weight of the case upon his extended hands, in an attempt, as he stood below it, to prevent its slipping. Immediately he realized that his left wrist had been injured. I saw him on the following day and determined the presence of a fracture of the inferior extremity of the left radius ; the lower fragment was slightly displaced backwards ; the line of fracture upon the outside was apparently one inch above the tip of the styloid process of the radius, whence it ran transversely less than half an inch above the articular surface; the relations of the ulna were unaffected. This case was also examined by Drs. F. H. Stuart and E. H. Bartley.

An additional case, unusually free from any possibility of the presence of any other force than that of over-extension, is narrated by Prof. Mac-Leod, of Glasgow, in the *British Medical Journal* of July 12th, 1879. p. 39. The case was this: " A young man contended with an older and stronger man in a test of strength by placing elbows on a table, interlocking fingers, and then pressing back against each other palm to palm. The hand of the young man became violently extended, until finally something gave way with development of sharp pain in the radius, with the well marked deformity characteristic of fracture of the inferior extremity of that bone."

Fractures of the lower extremity of the radius have resulted from falls upon the back of the hand. Hamilton relates an instance of the kind.

Gordon relates an instance in which a man, aged twenty, whilst leading a horse out of a stable, had his fore-arm compressed between the door frame and the side of the animal, the hand being flexed at the time. The radius was broken about an inch and a half above its lower end.

In fine, the character of the violence by which fractures of the lower extremity of the radius are produced is fourfold, viz. :

1. Falls upon the palm of the hand; common.

2. Simple over-extension of the hand; unfrequent.
3. Falls upon the back of the hand; rare.
4. Simple over-flexion of the hand; very rare.

CONCLUSIONS.

After this analysis of the facts involved in fractures of the lower extremity of the radius, we may, with greater facility and accuracy, hope to trace the sequence of the causes to which may be due the varying conditions which different fractures present.

What is the order of events which happen in falls upon the palm of the hand ?

First, the impact is sustained by the thenar and hypothenar eminences over the base of the metacarpal bones. In extreme extension the prominences of the trapezium and unciform bones became involved in the impact. That the fall should come upon the palm of the hand at all, it is necessary that the hand should be previously somewhat extended, in which movement the points where the impact is to be sustained are carried back of the axis of the radius.

The immediate effect of the impact is to increase this extension. The distal row of the carpus glides backward upon the proximal row, the latter glides forward beneath the articular surface of the radius until it is checked by the distended anterior ligament, and the radius rests upon its posterior surface. The flexor tendons also—if the muscles be in contraction—antagonize this extension and reinforce the anterior ligaments. However, if the flexor muscles have been sufficiently strongly contracted to resist the further tendency to extension at the moment of the impact, the hand and carpus will be driven *en masse* upwards, as the resultant of the forces of extension and flexion, and the narrow convex surface of the upper row of the carpus will impinge against the broad shallow concave articular surface of the radius. In the transmission of the force to this point it will have already been greatly decomposed. The elastic adipose, fibres and muscular cushions of the palm, and the two rows of carpal bones, cancellous in structure with their interposed cartilages, unite in repelling shock; whatever of strain has been put upon the flexor muscles is just so much force transmitted to the origin of the muscles and there expended; whatever of ligamentous strain is sustained, is just so much of the original force transmitted to the insertions of the ligaments and there expended; whatever of the force of the original impact which is not thus expended is finally received upon the articular surface of the radius. Here an admirable arrangement is found to break up and render harmless whatever of force yet remains: a buffer of cartilage upon which rests a cone of cancellous tissue which is grasped above by a cylinder of strong, thick compact walls. All force transmitted to the lower extremity, other than that of very great intensity, of the radius, is here finally decomposed and repelled. If, however, as the result of very great violence sustained, as in falls from a height, an impact of the carpus against the radius is occasioned more powerful than can be resisted by the mechanism described, what will be the result? The problem is a simple one. The carpus is a blunt wedge; driven against the concave articular surface of the radius with sufficient force it will split it longitudinally. There is no possibility of a cross-breaking strain being exerted, and if in any case transverse lines of fracture are found, they must have been occasioned by some other mechanism.

The description of the results of falls upon the palm of the hand applies equally well to those of falls upon the back of the hand, substituting the words flexion for extension, extensors for flexors, and noting that the line of force falls in front of the axis of the radius instead of behind it. As a matter of fact, however, falls upon the back of the hand are of very rare occurrence.

The existence of these longitudinal lines of fracture in certain cases of injury from falls upon the palm of the hand was first signaled by Dupuytren fifty years ago, who, never permitting an opportunity to gain exact knowledge by an autopsy, stated that where an autopsy had permitted an examination of a recent fracture of the lower extremity of the radius, he had often found the extremity shivered with radiating lines of fracture, as if it had been struck a blow with a hammer. (*La Lancette*, VI., No. 4; quoted by Voillemier, *Arch. Gen. de Med.*, 1842, LVIII., *p.* 263.) But Dupuytren made no attempt to differentiate the causes of this shivering from that of the transverse fracture; the injury as a whole was attributed to the direct transmission of shock to the radius through the carpus. Nelaton, in his *Éléments de Pathol. Chirurg.*, I., *p.* 740, treating on the pathology of fractures of the inferior extremity of the radius, bases his views on the results of certain experiments on the cadaver; in these, by striking forcibly upon the upper end of the bones of the forearm, held vertically while the palm of the hand was supported upon a firm surface, he produced transverse fractures of the lower extremity of the radius. He imagined that the backward bending of the midcarpal articulation was so free that the first row of the carpus was permitted to impinge against the ground, and that through this directly the shock had been transmitted to the radius above. Upon the cadaver he secured transverse fractures only. *Post hoc, propter hoc!* Strong in his assumption of the relation of the carpus to the radius, he announced the formal doctrine that transverse fracture of the radius just above the lower articular surface is the result simply of the fact that the bone, having been caught between two forces, the resistance of the ground on the one hand, and the weight of the body, increased by the velocity of the fall, on the other, has given way at its weakest point. The only possible effect of the direct crowding of the first row of the carpus upward against the expanded concave articular surface of the radius, if it had taken place, was entirely overlooked by Nelaton, and longitudinal fractures had no place in his pathology. The genius of Nelaton was sufficient to impress his views on his compeers for a whole generation. Malgaigne and Voillemier accepted it, though both acknowledged the occasional production of the fracture by the unaided force of over extension of the wrist. The surgeons of England and America have accepted it, and in great measure do accept it to this day, and much ingenuity has been displayed in demonstrating that the usual point of fracture in the radius, where it commences to expand with thinning compact walls and abundant cancellous tissue, is the weakest part of the bone, while the necessary preliminary of explaining how the longitudinal force, assumed to be acting, was converted into a transverse strain has always been strangely overlooked.

A new fact, however, was added to our knowledge of these injuries by Dr. H. J. Bigelow, when, on January 11th, 1858, he presented to the Boston Society for Medical Improvement a radius, the lower articular surface of which presented a *star-shaped* crack, and corresponding fissures extending therefrom upward in the shaft for more than an inch without transverse fracture. Other injuries received at the same time had caused the patient's death. This was the second case of the kind which had been verified by autopsy in the experience of Dr. Bigelow.

Here were two cases of longitudinal fissure of the inferior extremity of the radius, uncomplicated by transverse fracture. By what mechanism had they been produced? If, as Dr. Bigelow thought, the bones of the wrist had acted as a wedge to spread the corresponding hollow of the articulating extremity of the radius in these cases, would not the tendency to produce the same effect have been present in all cases where the bones of the wrist were crowded up against that articular extremity? How, then, can transverse fractures of the bone be the result of a force which merely produces a violent impact of the carpus against the articular surface of the radius?

The conclusion is inevitable; there is no principle of mechanics, no peculiarity in the construction of the bone, no peculiarity of the dynamic conditions present which can alter it, that such an impact, as in these two cases, so in all, will, when sufficiently violent, produce longitudinal fracture, and that only.

In view of the special adaptation of the mechanism of the wrist joint to successfully resist force applied to the palm of the hand, it is evident that the original force must be one of great violence to produce longitudinal fracture. In injuries resulting from comparatively slight violence—which includes the great mass of recognized fractures of the lower extremity of the radius—longitudinal fracture cannot be present; but in falls from a height in which the weight of the body with great momentum comes upon the outstretched palm, the conditions are presented in which, if our reasoning be correct, longitudinal fracture may be possible. These are the cases likewise in which other severe injuries are likely to be present, which, by their fatal results, may make actual inspection of the fractured bone possible. These are they which find their way to hospital wards and dead houses. These formed the class of cases which Dupuytren inspected, and which gave rise to his graphic description of fragments shivered and splintered as if struck by a hammer. To this class belong the twelve cases immediately under examination, in which it has already been recorded that ten present the articular extremity of the radius, comminuted by lines of vertical fission.

But the claim may be made that the comminution of the lower fragment is the result of impaction and penetration therein of the upper fragment, an explanation that has heretofore been generally accepted. But of the comminuted specimens before us not one presents any indication of impaction of the upper in the lower fragment. In one the separation of the two fragments from each other is still incomplete, the fragments remain still unseparated, with their periosteal investment intact posteriorly, and the anterior surface presents a gaping transverse fissure, up to which ascends from the articular surface a vertical fissure. The appearances presented by the specimen from Dr. Hamilton's cabinet are especially notable. From a central point in the concave articular surface radiate four fissures, which ascend vertically for two or more inches. No fragments have been split off, the bone would have been simply fissured longitudinally, were it not that there has been a transverse fracture, which has traversed the vertical fissure, and produced a lower fragment which is composed of several pieces. The precedence of the vertical fissures to the transverse fracture is unmistakable, and the very shapes and relations of the pieces of the lower fragment negative absolutely their production by the driving of the upper into the lower fragment.

In yet another specimen, there is a distinct indentation in the radial articular surface, with a stellate crack radiating therefrom, one of the rays of which reaches the posterior margin of the bone, and is discernible as a distinct fissure for an inch vertically.

The facts with regard to longitudinal fracture, which have now been demonstrated, may be thus summarized:

1. Direct impact of the carpus against the articular surface of the

lower extremity of the radius tends to produce longitudinal fracture or fissure in the radius.

2. Longitudinal fracture or fissure only occurs where great original violence is inflicted.

3. Longitudinal fracture or fissure may occur alone, or may be complicated with transverse fracture or fissure.

4. Longitudinal fracture or fissure, when complicated with transverse, is produced first in point of time.

5. The force which produces longitudinal fracture or fissure in one case, will always tend to produce the same result, and will not sometimes produce the one and sometimes the other.

One hundred years have now elapsed since Pouteau first taught that injuries at the wrist from falls upon the palm of the hand, instead of being simply sprains, or incomplete dislocations, were fractures of the radius near its lower extremity; but I believe that I have the fortune of having now first recognized the full relation to these injuries of longitudinal fractures of the inferior extremity of the radius, and of elaborating their pathology. The pathology of Nelaton has completely blinded surgeons to the existence or true character of this injury. Bigelow, in 1858, had the clue which might have led to the establishment of the truth, but he did not pursue it. Two years later Lecompte made the most important advance in our knowledge of these injuries since the labors of Dupuytren, when in the *Archives Generales de Médecine* (1860, p. 653), he published his masterly and conclusive demonstration, that transverse fractures of the lower extremity of the radius were always due to *arrachement*, a force of avulsion, cross-breaking strain exerted through the anterior ligament of the wrist. Having this foundation which Lecompte has laid, it has been possible to add to it further true exact knowledge.

The discussion of longitudinal fractures of the inferior extremity of the radius had its origin in the hypothesis that in a given case the contraction of the flexor muscles was sufficient to overcome the tendency to extension which is impressed upon the hand and carpus as the immediate effect of a fall upon the palm of the hand. In cases where the excessive violence described as necessary to produce longitudinal fission fails to be inflicted, no evil effect beyond a certain amount of concussion is sustained by the structures of the wrist. But observation shows that in many cases, either the resistance of the flexors is overcome, or taken unawares they fail to contract at all, and the movement of extension going on, its force is brought to bear chiefly on the anterior radio-carpal ligament. The posterior surface of the first row of the carpus is now strongly pressed up into the posterior portion of the concavity of the articular surface of the radius, all pressure is removed from the anterior portion and is now wholly sustained by the posterior portion. The extended hand and carpus constitute a powerful lever, which, supported by the posterior wall of the radius as a fulcrum, exerts a resistless strain on the anterior ligament, and through it upon the anterior edge of the radial articular surface.

The result is either avulsion of that portion of the bone controlled by the ligament or rending of the ligament. The cross-breaking strain by which transverse fracture may be accomplished is thus produced. The full discussion of the mechanism of this force of avulsion, together with its demonstration upon the cadaver, and the development of certain new and important points in the pathology and treatment of the resulting injury were contained in my memoir of March, 1878. I shall not repeat it. There is but little which I then said that later experience prompts me to recall or qualify. I stated then that almost invariably in experimenting upon the cadaver by producing over-extension of the wrist, the lower end of the bone was torn off. After a larger number of experiments I would make that less strong, and say that as a rule the ligament remains intact and the radius is fractured transversely at a point rarely higher than a half inch above the articular surface. In some instances, however, the ligament gives way and the radius remains intact, and occasionally one of the carpal bones is fractured. The shape and extent of the lower fragment torn off in these experiments has varied considerably. I have been able to produce almost perfect fac-similes of the various forms of the lower fragment already described as presented by the specimens of recent ante-mortem fracture.

By these experiments upon the cadaver, I have been able to determine the reason for these variations.

Of the three fasciculi which compose the anterior radio-carpal ligament, the middle is the weakest, and, if rupture of the ligament takes place, it is here that it begins. The outer part of the external band, where it is reinforced by the external lateral ligament, is the strongest. Thus it may happen that these fibres alone may remain intact, and the tip only of the styloid process be torn off; or, the entire external band remain intact, and the styloid process, with more or less of the articular surface, be torn off. If both bands remain untorn, the most frequent condition, the whole of the articular extremity will be torn off and a nearly transverse line of fracture be produced.

I have thus passed in review all the facts known to me, which bear upon the explanation of the occurrence of fractures of the lower extremity of the radius, with a candid desire to establish scientific truth.

The facts of anatomy, the facts of mechanics, the facts of pathology, of clinical experience and of experiment, combine to form the facts of the case upon which judgment is to be formed. All alike negative the theory that the ordinary transverse fracture of the lower extremity of the radius is the result simply of the bone having been caught between two counter-forces, the point at issue between my commentator and myself. These facts, however, have given us in response to our questionings a fuller and more accurate knowledge of the real pathology of the varying conditions which different cases of this fracture have been found to present. The outcome of the whole must be an improved therapeutics. more rational adaptation of treatment to conditions, more correct practice, better results, and a diminution of the danger, suggested by my friend at the outset of his article, that any reputable surgeon should be unjustly ruined in reputation and prospect as the result of his treatment of a case of fracture of the lower extremity of the radius.

www.ingramcontent.com/pod-product-compliance
Lightning Source LLC
Chambersburg PA
CBHW022034190326
41519CB00010B/1717